Frances Fisher Wood

Infancy and Childhood

Frances Fisher Wood

Infancy and Childhood

ISBN/EAN: 9783337371463

Printed in Europe, USA, Canada, Australia, Japan

Cover: Foto ©berggeist007 / pixelio.de

More available books at **www.hansebooks.com**

INFANCY AND CHILDHOOD

BY

FRANCES FISHER WOOD

NEW YORK AND LONDON
HARPER & BROTHERS PUBLISHERS
1897

CONTENTS

CHAP.		PAGE
I.	Preventable Diseases	1
II.	The Young Babe	12
III.	Regularity in Feeding	24
IV.	System in Sleeping	31
V.	Rational Dress	45
VI.	Digestive Disorders	53
VII.	Sense-development	60
VIII.	Rational Feeding	66
IX.	Sterilized Milk	75
X.	From Infancy to Childhood	92
XI.	Normal Obliquities	102
XII.	Value of Milk as Food	110
XIII.	Contagious Diseases	119
XIV.	Variation of Rules	128
XV.	The Nursery	132
XVI.	To Avoid Self-consciousness	139
XVII.	The Nursery-Maid	146

INFANCY AND CHILDHOOD

INFANCY AND CHILDHOOD

CHAPTER I

PREVENTABLE DISEASES

Sir Joseph Fayrer reports that in England preventable diseases kill 125,000 persons a year, and entail a loss of labor from sickness estimated at $40,000,000 per annum. The Prince of Wales, when presiding at the Hygienic Congress of August, 1891, said: " If these diseases are preventable, why are they not prevented?" Statistics showing a steady decrease in the relative number of deaths prove that each year a larger proportion is prevented. In England the death-rate in 1660 was 80 per thousand; in 1690 it fell to

42 per thousand; in 1750 to 35 per thousand; in 1850 to 25; and in 1890 to 18 per thousand. That is, in England the ratio of deaths to population was nearly five times as great in 1660 as it was in 1890. The average duration of human life has doubled since the time of the Roman Empire, and to-day the mortality from all diseases is in Italy as great as it was in England a century ago.

The greatest decrease in the death-rate was in the seventeenth century. Then Europe, emerging from the superstition of the dark ages, first conceived the possibility of natural causes for disease and death. That era marked, in England at least, the beginning of public sanitation. People began to entertain the idea that cleaning the streets was more effective to stop the plague than the prayers which they had been wont to offer to the images of their patron saints.

The second great advance in the prevention of disease has been inaugurated within the past twenty years. The faint conception of the seventeenth century, that disease might have a natural cause, culminated in the now recent discovery of what that cause really was. With the acceptation of the germ-origin of disease, public sanitation and medical science began to pass from indefinite, general measures for the prevention of disease to more and more definite and specific safeguards; and the public health of the past decade has been indicated by further decided and still increasing diminution of the death-rate in all civilized countries.

That preventable diseases are still in great measure not prevented is due to the fact that the vitally important and growing knowledge of germ-infection, and the great laws under which these factors are operative (which is reducing the practice of medicine from a blind art to an exact science), are not as yet accepted in practice except by the better educated members of the medical profession. Even in their hands it is limited in usefulness because not supplemented by similar knowledge on the part of the patients whom they attend. A physician is not invited or permitted to prevent disease in the families which he attends. He is usually not called until acute disease has already fastened on his patient. He is consulted about the cure, and not called for the prevention of disease.

A certain class who insist on being facetious, even when facing death, add to the current stock of feeble jokes about the physician's self-seeking, the assertion that preventable diseases are not prevented because such a course would be suicidal to the physician himself, limiting his income, and finally depleting the ranks of the profession. While it is true that with the decrease of disease a smaller percentage of men will be needed in the medical profession, such decrease, unfortunately, must be too gradual to affect men already in practice, and in future can operate only to call fewer and better men to the medical profession. To these men the change of work from the cure to the prevention of disease cannot but be wholly beneficial. Their ranks once purged of charlatans and ignorant pretenders, who thrive on the superstitious fears of patients in the grasp of acute and chronic disease, the profession would be greatly raised in the general estimation, and its work would become in reality vastly more important to the community.

The very idea of the prevention of disease presupposes a comparatively large body of skilled and scientific medical advisers—men who prate not of symptoms, of potencies, and of cures, but who, in cases under their watchful eyes, can anticipate and therefore prevent acute illness. A physician will not then be called in an emergency (merely as a last desperate resort), but will be consulted regularly by every family, not about scarlet-fever and diphtheria, but as to diet and dress, exercise and education, and will be called upon to adjust all the vital details of daily life that make sound health. Not only will many men be needed, but their work will be less exhausting and more satisfactory to themselves and to the community than under the present system.

A true physician unites with the brain of the scientist the heart of the humanitarian. In the prevention of disease he, as a scientist, feels all the mental exhilaration of conquest, while as a humanitarian he experiences the highest degree of happiness in being able to confer the greatest possible benefit upon present and future generations.

The world's work in the future prevention of disease must come under three great heads. The first of these in point of time, and possibly of importance, lies in scientific discovery by original investigators in the fields of biology and hygiene. The second lies in the appropriation of these results by the medical profession, and in their practical application in every possible direction. The third lies in an extension of the knowledge possessed by these two classes (and at present almost limited to them) to the community at large, especially to the mothers, without whose co-operation a physician is powerless to enforce his principles.

The purpose of the present work is to assist in the prevention of disease in this last direction. It makes no claim to original scientific investigation; and for the physician so much which is technical and exact has already been written upon the same subject that he has no need of popular works bearing upon it. But the books which are so valuable to professional men are too abstract and technical, or too voluminous, for the general public. There is hence an urgent demand for reliable information, simply presented, which may be easily comprehended by unscientific people, especially by mothers; they, when it is brought to their knowledge that much of the disease current among their children may be prevented, earnestly desire to learn how to begin working towards that prevention.

Such information is not intended to make the mother independent of the physician, but rather to lead her to appreciate the fact that only the soundest medical advice, supplementing her most intelligent and earnest efforts, can diminish the amount of illness prevalent in her family. The slight knowledge she may gain cannot make her independent of the scientific skill of her physician, but is necessary merely that she may be able to co-operate with him to produce the desired immunity from illness.

That the mother's influence is so important an element in the prevention of disease arises from the fact that most of the successful work in that line must be done with children under five; and that after fifteen years of age it is not only too late to begin, but almost too late to produce any appreciable and permanently favorable effect upon the general constitution, by that time thoroughly established in disease tendencies. After puberty the greater part of the work done by the physician must be in the treatment rather than for the prevention of illness. The size, general health, physical strength, nervous force, and power of endurance of any child is practically determined before he is five years old. Indeed, it is determined in a great measure before he is born; but during the first years of life the susceptibility to improvement is at its maximum, and hereditary strength may then be increased or hereditary weakness modified to a degree that is never afterwards possible. The united power of the mother and of the physician to influence the future physical condition and vitality of a child is practically limited only by the general public ignorance which exposes a child to outside danger and infection, from which it has been with much effort successfully guarded at home.

Here, too, in public hygiene the scope of woman's influence is increasing, and it is to be hoped that, once educated to a conception of the hygienic conditions that should prevail in the home, she will widen her influence to extend and improve general legislative measures for the protection of the general public. She can thus not only guard her own children, but at the same time protect the children of mothers less accurately informed of the dangers of infection.

The time must come when contagious disease will be considered a crime, either of the individual or of the community by whose carelessness it is propagated. During the last century the responsibility for disease was universally placed upon the Almighty, who was thought to send it either as a punishment or as a warning of sin. During the present century we have transferred the responsibility largely from the Creator to his creature, the microbe. It is to be hoped that the coming twentieth century will convince mankind of the error of both these views, and that we may learn to place the burden upon neither the infinite nor the infinitesimal, but to realize that it is too petty for the one as it is too vast for the other, being of exact size to fit the shoulders of the genus homo. When we as individuals and as members of the community accept the full personal responsibility for all diseases and deaths except those resulting from accident and old age, we shall have gone a long way towards general physical regeneration.

CHAPTER II

THE YOUNG BABE

THE care of the new-born child is really less difficult than at first would appear. It is quite as important to know what not to do as what to do for its comfort. Most mothers and nurses try to do too much, directing their efforts, however, in lines where they are not only wasted, but are positively injurious to the feeble recipient. A young babe is an uncomplicated being; the latent, intricate organization of adult life is with it still inactive. When a guest of maturer years enters our household we need, in order to make him comfortable, to learn his habits, his tastes, his prejudices. But an infant comes to us without any habits or tastes or prejudices. It has eyes, but they see not;

ears, but they hear not; tongue, but it questions not; hands, but they meddle not; feet, but they go not into mischief.

The new-born child has but two sets of organs in conspicuously active operation—the respiratory and the digestive. As the lungs work independently of any outside aid, we may be allowed to consider a child during the first few months of its life as practically an incorporate stomach; and we shall be safe in limiting our contributions in the way of attention to that organ. The majority of mothers consider this view a base libel, and are not slow to proclaim that every mother's child of them all is from the very first manifestly more than a stomach; that every child loves to sit up and look around, to see visitors and to be dandled; that it shows its will-power in its determination to be carried, held, or rocked; that it knows full well when it desires to be fed, and equally well when it desires not to sleep. But all this is, on the face of it, really absurd. An animal, human or brute, is born with but one instinct fully developed—that of hunger. Whatever else a very young child knows or desires has been artificially communicated by the attendants, who have forced upon it taste and desires unsuited to its immaturity. If you would know at what age a child should begin to sit up, wait until natural, healthy development prompts it to sit. It will creep when the proper time comes, and walk when its limbs are sufficiently strong, and the power of equilibrium is adequately established in its feeble brain. If a child is surrounded with the proper degree of light, warmth, and air, and if its stomach is properly taken care of, we may rest assured nature will do the rest.

A child should literally be intelligently let alone. It should not be handled, or held, or rocked, or amused, nor should its attention be attracted in any way. For the first five or six months it should lie quietly in its bed, or basket, be regularly fed, and as regularly encouraged to sleep. It will of course get tired. Therefore it needs occasional turning, with change of position, and a gentle rubbing of the limbs or back. A good rule is to stroke the little body for a few minutes, and to change its position every time the baby needs to be made dry. The natural rapid growth of infancy makes the flesh tingle and the limbs ache, and frequent rubbing with the palm of the hand promotes future health as well as present comfort.

To many mothers it seems impossible to follow this simple rule of leaving a child alone, to be influenced by favoring natural conditions. To such it is only necessary to say that it has been successfully accomplished in many cases even in America, where babies are seemingly unusually troublesome, while such abstinence from meddling is the accepted rule in many other countries. That American babies seem to need so much diversion and entertainment cannot be ascribed to the climate, for the Indian pappooses are invariably left to themselves, except when dressed or fed. Nor can it be a race characteristic, for in England it is considered very injurious to permit a child to sit —that is, to bear its weight upon its spine— before it is six months old. The true cause probably lies not in the children at all, but in the recognized restlessness and nervousness of American mothers, who, by expending their fiercest energy in injudicious attention to their children, perpetuate in the coming generations this nervous tendency, so highly undesirable, because ever fatal to the highest physical vigor and the best mental development. Many a child languishes for lack of the few things that are really necessary to be done, but that are neglected in favor of unsuitable attentions lavished upon it only to its detriment.

In order to preserve for a young babe the proper conditions of light, warmth, and air, and yet to lift and carry it as little as possible, it is necessary to have for its first nest a movable bed. Any basket with the sides and bottom carefully protected and padded will serve, but the most convenient is the regular dog-basket, with a hood on one side. This, when properly draped, serves to exclude draughts, while the drapery may easily be readjusted to vary the degree of light. If a child occupies a stationary crib, it must be moved from its bed whenever its room is aired or cleaned, or is needed for other purposes. But when such a basket is used, the child and bed together may be changed from one room to another, or from one part of the room to a darker or lighter corner, or to a cooler or warmer one, as convenience or comfort may suggest. Most important of all, a mother, without confining herself to the nursery, can keep the infant under her own eye while engaged in her ordinary daily occupations. Even though she does not personally feed and care for her baby, she can thus superintend and criticise the nurse's efforts.

By this method she may also experience the greatest of all maternal enjoyment—that derived from watching the daily development of her child. Also, she can at the same time, without interruption or fatigue, conveniently sew or read, write or study, receive visits, or direct her household affairs. Outside work, either of a social or a business nature, should be undertaken by no woman during the first six months of her infant's life. She herself needs, and has by the rights of maternity earned, that amount of rest from either nervous or physical overstrain. Furthermore, years of a mother's devotion later in life can never compensate a child for neglect during the first few months.

After the sixth month a child usually begins to teethe. Voluntary muscular action is then more frequent. Feeble beginnings of individual will-power are manifested. The babe gradually recognizes the world outside of itself. It is no longer merely an animated stomach; other faculties and functions start into activity.

All this varied development makes increasing demands upon the nervous system, reacting upon the physical nature, and immediately manifesting themselves in a checking of the phenomenally rapid growth to be noted during the first six months of every healthy child's existence. If the precious first six months have been properly used, the development of the second six months is not less rapid, although it expends itself in other directions than in purely physical growth. This, however, should normally take place, without any disturbing elements or violent check. But if the first half-year has not been employed to build up the maximum of physical strength, and to train the child into normal, healthful habits, the second half-year is confusion worse confounded, and in too many cases records the death of the child.

Habit rules us all, but is absolute master of the unresisting infant. A baby is a natural autocrat, recognizing no authority. It is in vain that the mother tries to induce it to sleep at the proper time, or strives to hush its cries when the desired food is not forthcoming. What she cannot accomplish, however, the simple power of habit can bring about without a struggle. If the child is fed at absolutely regular periods, it will be hungry then and at no other time. If it sleeps day-by-day by the clock, sleep it must when the hour strikes, whether it will or no. Even the stomach can be trained into the habit of digesting the maximum amount of food necessary for the full nutrition and growth of the body; and when so trained, it possesses marvellous power to carry on its accustomed work under such temporary irritation or derangement of the general system as would render a child with a weak stomach seriously if not violently ill.

The first six months of life, therefore, form the mother's golden opportunity. If she do not then lay well the foundation, the whole superstructure must betray this primary defect. Then and then only will all the elementary forces of nature be on her side. Later some elements, if not all, will be against her. If she neglect the child at first, or leave it to the untrained care of a nurse, she will, as a penalty, certainly spend many times six months during its later life in nursing it in illness or caring for it in invalidism. Complaints of even the moderate expenditure of money and time necessary to establish an infant in sound health during the first year of existence come from many parents who ungrudgingly give a hundred times as much to the same child in its twentieth year of life, and this because they fail to realize that that first year was really more important in determining the success of its future life than the whole nineteen which followed. What we must do, we can do. If a woman is seriously ill, she can and does lay aside her regular occupations. If she goes abroad for a half-year, she returns to find that neither

her home nor the American continent has been revolutionized during her absence. If a mother recognized the real necessity for such a sacrifice, she would find that she could easily give six months of superintendence to the early life of the child, especially since she need not by so doing relinquish the quieter, less exhausting occupations that may be carried on inside her home, and must sacrifice only such social excitement or heavy professional duties as exhaust her strength and rob the child of her care.

The emphasized importance of the first six months of life arises not from the fact that there are any unusual dangers to be expected during that period, for in the ordinary course of events diseases culminate rather in the second half-year. But the first six months are the most valuable in any scheme for the prevention of disease. Precisely during this period is laid the foundation for the habits of digestion and sleep, which, if well established, will carry a child safely through the remainder of the first year, and will also influence for good the whole of his subsequent life. The majority of children seem to flourish until they begin to teethe; then the slight irritation and physical disturbance of this process reveal any original weakness of the constitution—a weakness in many cases increased by unscientific care during the earlier period. Differing from the popular opinion, all the best medical authorities agree that "teething" never causes disease; but it often betrays weaknesses hitherto unsuspected, which have resulted from poor feeding, over-excitement, or improper clothing, and which existed before and up to the time when teething began.

CHAPTER III

REGULARITY IN FEEDING

The popular notion that the second summer is the most precarious period of a child's existence is absolutely disproved by all statistics. Four times as many children die in the first year of life as in the second, and with each month's existence a normal child's hold on life is strengthened. If a baby is steadily deteriorating in physical condition, the second summer is, of course, more dangerous than the first, but if, as the result of proper care, the child is following in all its powers and proportions the more natural rule of accelerated growth manifested throughout the whole animal kingdom, the second summer finds its chances of life and of freedom from disease very much more favorable than they were during the first year. During the second half of the first year a mother should in no wise relax her vigilance in the care of her baby; although if the first half has been judiciously used, the increased vitality, the greater powers of endurance, and the establishment of the regular habit of sleep will permit her to maintain an equal degree of caution while devoting to the work less time and effort.

Teething is a natural operation, and ought to take place in a child as easily and as normally as in other young animals, which at worst suffer but a few days' irritation of the gums, and refuse a few of the usual feedings rather than endure the discomfort caused by mastication. If a baby is accustomed to awake at all hours of the night, and to sleep at any hour of the daytime, then even the slight irritation of teething interrupts or altogether prevents its sleeping. But if it has during several months slept soundly all night and napped regularly each day, the habit will have become too strong to be broken by any slight pain. In one case the restless child grows weaker day by day for want of adequate sleep; in the other, the rested child, even though teething, grows stronger every day, because still enjoying its regular amount of sleep.

The same principle applies to food. A child that is injudiciously fed throughout the night, and given nourishment irregularly whenever it cries during the day, suffers in consequence from impairment of its digestive powers, and becomes susceptible to germ infection throughout the mucous lining of the alimentary canal. Such a child, while it suffers from chronic hunger caused by the craving of the general system for adequate nourishment, yet never at any one time experiences the wholesome acute hunger of normal health, and therefore, when teething, alternately refuses food or consumes it so voraciously that each feeding only serves to aggravate the stomach trouble already existing.

Its flesh, already too white and soft from malnutrition, now wastes away so readily as to convince the mother more firmly than ever that teething is a "disease." The diseased condition in reality existed previously, and the soft white fat was, to more experienced eyes, one of its manifestations. When a really well-fed, well-nourished child begins to teethe, habit induces it to take a due amount of food, even though the nipple does hurt the sore gum, or the piece of bread does touch the tender spot over the pricking tooth. The child, in short, must eat, because its hunger is so acute and habit so strong.

My own baby, reared in the routine here advised, had twelve teeth when a year old, sixteen teeth at sixteen months, and the entire set at two years of age. During that time he refused not a single meal, and lost no night of sleep. His first three double teeth were through before it was suspected that he was cutting one. Such a record is the rule rather than the exception with children who have been reared by the proper system.

With this explanation, the emphasis laid on the first six months of life will not seem unduly exaggerated, and the details of the care necessary to produce such results as described will seem to mothers to be of sufficient importance to be dwelt upon at length.

As the new-born babe has no regular established habits, the operations of sleeping and eating at first naturally tend to interfere with one another. When the child should feed, it is often asleep; and when it should sleep, it persists in being hungry. The almost universal injunction given in medical works is never to interrupt a child's sleep for any purpose whatsoever, not even for feeding. But this rule, if strictly carried out, absolutely precludes any regularity either in sleeping or eating. An infant does not distinguish night from day. It will quite probably sleep four or five hours at mid-day, and desire to remain wide awake and demand frequent feeding all night. Both for the comfort of the mother and the health of the child it is of course necessary that the latter should learn the difference between night and day, and be induced to sleep at the proper time.

This lesson can best be taught by indirection. The question of feeding should, despite the verdict of the learned doctors, have first consideration. Throughout the animal kingdom the necessity for food is the primal and all-absorbing instinct. From the very first a child should be fed equal quantities of food at absolutely regular periods, even though he must be awakened from sound sleep to receive this amount. And from the beginning a child's stomach should enjoy at night an interval of rest, even if this interval is at first not longer than five hours. If the mother begins when her babe is young enough, it requires but a few days to establish this useful habit. Those few days will be days of weariness to the mother and distress to the child, but that first victory will render all the other subsequent educational battles easier. One such successful contest may, indeed, become also the basis of a sound system of future discipline, the effects of which will endure as long as the child lives.

Many important advantages grow out of this first rational habit of regular meals when it is once firmly established. The baby is no longer found asleep when it should eat. Periodic hunger, resulting from a regular habit of feeding, will soon waken him at the proper time. His stomach, accustoming itself to the new routine, demands food at the usual time, and is not clamorous at any other period. Since it is not overfed at one time and underfed at another, the stomach learns to desire and to assimilate all the food required by the system, and thus the body accomplishes the maximum amount of growth.

CHAPTER IV

SYSTEM IN SLEEPING

It is not wise to attempt to teach a baby more than one thing at a time. Neither the mother's strength nor the child's endurance should be taxed to carry on several contests at once. All forces may wisely be concentrated on the vital points, one after another.

It will require some days after the baby submits to regular feeding to ascertain just how much it can assimilate each time it eats, and what is the best interval to maintain between meals. During this experimental period, even though the hours of sleep do not receive especial attention, the hours of feeding will of necessity determine to some extent the periods of rest. If sufficiently well fed, a baby even a month old can easily go without food from 11 P.M. to 5 A.M. And when once convinced that no food will, in any event, be forthcoming between these hours, it will usually fall into the habit of sleeping soundly during this time. Should it prove impossible, even after weeks of patient experiment, to induce the child to sleep these six hours without food, it may be considered as positively proven that the food it consumes during the day is not sufficient for the demands of its system. If it has been fed regularly every two hours during the day from 5 A.M. to 11 P.M., it has had ten meals, and this should be enough to carry any child through the other six hours of fasting, unless the food is essentially deficient in quantity or quality, or both. Therefore, if after persistent trial the baby cannot be made to sleep at night without feeding, its food should at once be the subject of consultation between the mother and her physician. As almost all young children are fed too much in quantity, the quality of the food is, in ninety cases out of a hundred, the element requiring investigation.

If, however, this unbroken night's sleep is once accepted as a rule by the child, we should begin at once gradually to lengthen the time at both ends. The child should not be fed a minute before five o'clock, even if it is awake, and when it does not awake on time it should be allowed to sleep on as long as it will. In this way it will itself, gradually and without any struggle, increase the hours of rest.

The daily rule for sleeping and eating for the average child is that it should, when one month old, be fed every two hours from 5 A.M. to 11 P.M., sleeping from that time until 5 A.M. again. At three months it should be fed every two and one-half hours from 5 A.M. to 10 P.M., sleeping from 10 P.M. to 6 A.M. At six months it should be fed every three hours from 6 A.M. to 9 P.M., sleeping from 9 P.M. to 6 A.M. At one year of age a child should be fed at seven, ten, two, six, and nine o'clock. The first and last meals should be given to the child in bed, from the bottle, while the other three meals should be fed from bowl and spoon, in order to begin the weaning process.

During the fast of the night there should be always ready by the bedside a thoroughly clean nursing-bottle filled with water that has been boiled. If the baby is wakeful, fretful, or hungry, allow him to nurse from this. A few swallows will suffice to calm him. The ordinary heat of the chamber will render the water warm enough for a child in health. If the infant is delicate or ill, the drinking-water must be warmed to 98° Fahrenheit in a cup of water placed over an alcohol-lamp on the table. Sometimes when a baby is breast-fed it will not drink even water from a nursing-bottle, in which case it is necessary to moisten its mouth as often as it cries with a fine, soft, white cloth saturated with water. An older child should be fed with water from a spoon. Water the child must have, and in abundance, during the troublesome nights, when the habit of sleep is not yet established and the desire for night meals is not thoroughly overcome.

Besides this most important sleep at night, regular day naps must be established as soon as possible. A new-born babe should, and usually does, sleep most of the time; but if it is of a nervous temperament, there is danger, as it grows older, that it will fall into the habit of catching short naps at odd moments, and of indulging in no profound, lasting rest during the whole day. By the time a child is three months old it should have formed the habit of sleeping from ten to twelve hours at night, and of napping at least two to two and one-half hours twice during the day. The intervals after the 10 A.M. and 3 P.M. feedings are the most favorable times for these rests.

In my experience with my own baby this habit of regular day naps was the most difficult to establish. It required only three or four days to accustom the child to regular feeding, and but two nights' struggle to teach him to submit to the inevitable and to agree to consider the night as the time for sleep. But it was only after several weeks of unremitting effort that any regularity in day naps was established. By adhering to the first principle of leaving a baby quietly in its bed, one is debarred from the usual method of rocking or singing the child to sleep. Whatever lessons were taught had to be imparted as he lay in his basket. The child would sometimes fall asleep not more than half an hour before the time for the bottle, and when awakened by the periodic hunger, would have had insufficient rest; or he would fall asleep while nursing before he had taken the requisite amount, and must needs be aroused to finish the meal. We were finally obliged, whenever he was fed, to resort to the expedient of deliberately keeping him awake until he had consumed the usual amount, and at other times, as far as possible, to prevent his napping between the regular hours of sleep. When the hour for a nap arrived he was given his usual bottle, after which he was quietly and soothingly stroked down his back, sides, and limbs, and then turned over to lie on his stomach. The inside of the basket was darkened by the adjustment of the drapery, quiet was enforced in the sleeping-room, and he was left to sleep — to which the bottle, the rubbing, and the comfortable position soon wooed him. It was from the first easy to induce him to sleep, but difficult to prolong the sleep for any considerable period. By leaving him on his face during the length of time desirable for him to sleep, and saying "Sh!" whenever he awoke or called out, he in time imbibed the idea that it was sleep or nothing during those hours, and therefore he yielded again to the inevitable.

It was rather a weary time, that period of discipline, for no nurse could be trusted to exercise the degree of firmness and gentleness necessary to obtain the desired result without injury to the child. Yet it certainly proved time well spent; for after he was ten weeks old one had only to turn him over on his face when the hour for sleep arrived, and leave him entirely alone. It was absolutely certain that he would be asleep in five minutes, and would sleep for a fixed length of time.

For every hour spent in the initial training, innumerable hours of labor which would later be consumed in rocking the child to sleep are saved to the mother, to say nothing of the gain to the child in a nervous way—of dropping asleep quietly, resting so profoundly, and sleeping so long.

Our great scientists tell us that with all the superficial differences between the civilized and savage man only one divergence is vital—the savage thinks and plans for the present, the civilized man thinks and plans for the future. The mothers of the past who resorted to any expedient, however irrational, which rendered the work of the present moment easier, reverted to the methods of the savage. The scientific mother of to-day, who takes present pains in order to avoid future trouble, who increases the labor of to-day in order to diminish the sum total of effort necessary for the training of her child, thereby marks her system as one in harmony with the highest type of civilization.

A strict method of discipline is of course possible only with a child who is comparatively well. Rules and regulations must be held in abeyance through any severe illness; but the nearer we can approximate to regular habits, even with a sick child, the better it is for the child. By establishing strict daily rules and by maintaining a wholesome system of nourishment, sickness will be the rare exception for a young child, and while ill health may be allowed to modify regular rules, it need not abolish them.

The weak point in any system of home discipline lies with the parents. It has been said of Herbert Spencer's theory of education that it would be absolutely perfect if only the parents were perfect. Yet a good system feebly enforced is, in so far as it is enforced at all, certainly superior to a poor system. Twice when lecturing upon the training of children to an audience of women, I have, after the lecture, been approached by one of my hearers, who in each case made almost the same criticism upon what had been said. Though the two women were unknown to each other and in cities far apart, each said virtually the same thing. Both suggested politely that however admirable the lecture might be in its general scope, it could not be valuable to more than a small portion of the audience, since it was addressed to those only who possessed indomitable will power; and that while all would probably acknowledge the wisdom of my suggestions, few could be steadfast enough to follow them out against inevitable opposition. To which stricture I can, in connection with the present subject, say that if a mother clearly recognizes her lack of will-power, never is the time more propitious for her to begin to exercise what little she has than with the infant a few weeks old who has none at all. Perchance by that exercise her own too feeble will-power may be induced to keep pace with the growth of that of her child. In any event, it is her only hope for supremacy, and, therefore, is an opportunity that should be eagerly embraced.

Sound, restful sleep, both by night and by day, is more easily induced if from the first the child be taught to lie on its stomach and face. The only necessary precaution against suffocation is the provision of a smooth, flat, somewhat hard hair mattress without a pillow. The advantages of this position are many. Some one has said that half the diseases of infancy result from keeping the stomach too cold, and the other half from overheating the spine. By adopting the position suggested as the uniform one during the hours of sleep, the stomach and abdomen are kept so warm as to prevent colic and stomachache, and materially to aid the digestive process, while the spine and back of the head are no longer overheated by the increased temperature of the sleeping child. It may be a coincidence merely, but it is at least a significant one, that all the children the writer has known to rest habitually face downward have been unusually sound sleepers, and have enjoyed more than average good health.

It is surprising to see how early a child will discriminate and show preference for the face position, and how readily it accommodates itself to this attitude. A child from eight to ten weeks old will already have learned to turn its head from side to side to obtain the relief of a change of position.

A young baby on its back is as helpless as a turtle in the same position; its one possible motion is the throwing out of legs and arms, and each such movement uncovers the child and exposes it to draughts. Placed on its face, a babe two or three months old will not only rest itself by frequent changes of the position of all portions of the body, but, since it is powerless to reverse itself, it cannot get uncovered nor lapse into any unwholesome cramped position. It is quite otherwise when the infant is lying flat on its back. This position not only invites indigestion, but it also causes bad dreams and night frights, and promotes the dangerous habit of mouth-breathing.

The first basket for a child should be made up with but one sheet, which will serve to cover and protect the mattress. Over this the child should lie between woollen covers. The ideal bedclothes for a baby are small camel's-hair blankets, which weigh almost nothing, and yet are sufficient, one under and one over the child, for even the coldest weather. A small square of heavy double-faced Canton flannel laid under the child, between the night-dress and napkin, will prevent any wetting of the under blanket.

CHAPTER V

RATIONAL DRESS

To be ideally comfortable and well, a child should, during the first year of life, be clothed entirely in silk and wool. Knitted silk shirts in summer and wool in winter, with socks of the same material, make, with the napkin, one complete cover for the little body. Harsh, heavy, or coarse flannels should never be placed next the delicate skin of a young baby. The underwear that an adult finds grateful for its pure woolly roughness may so irritate an infant as to induce serious nervous trouble. We may now, however, obtain at moderate cost dainty knitted woollen shirts or flannel stockinet of such exquisite texture as to feel soft to even the rose-leaf delicacy of a new-born baby's skin. The garments that are worn over this should be made in princesse style, now known as the Gertrude pattern, without bands or strings, and buttoned behind, so that they can all be put on together. The inside dress should be of wool. Canton flannel as a material for infants' clothes is altogether an abomination. It is heavy and stiff and thick, but never warm. Cambric skirts and waists are entirely unnecessary, and, in such degree as they add weight and bulk, are really injurious. The garment worn immediately under the dress may be of silk-warped flannel, which will answer the requirements of warmth and yet will not show too deep a yellow through the thin dress. The dress itself, for comfort as well as beauty, may wisely be made of white China silk.

For a young baby, in cold weather, two garments are needed between the under-shirt and the dress. These should be made, one of Jaeger white stockinet, and the other of silk-warped flannel. Neither one should be more than long enough to cover the feet. These materials are so beautiful that they will require no embroidery or trimming. Simple feather-stitching will be sufficient to render both garments fit for a princess; and yet they will not cost as much and will be more durable than the usual long, heavily embroidered flannel skirt, and the longer, much-betrimmed cambric abomination called an over-skirt. Properly apparelled in the silk and woollen clothing, a baby has every garment as soft and warm as his own delicate flesh, and cannot be irritated or hampered by his dress, at least.

Silk-warped flannel skirts and white China silk dresses have an extravagant sound, and undoubtedly seem quite beyond the purse of many, who yet really spend double the amount that would be needed to purchase these articles on garments that are at once inartistic and unhealthful. The layette usually provided for a child is a barbarism. It is elaborate, yet not beautiful; expensive, but not useful; troublesome to make and keep in repair, and yet not comfortable for the wearer.

White China silk costs from fifty cents to a dollar a yard. The dress, like the flannel under-garments, may be made entirely plain, and, at most, should not be more than forty inches long. The expense of such a dress is not more than half that of the ordinary hideous over-embroidered gown, which is beyond home skill to make or home talent to launder.

On the subject of napkins a word remains to be said. The most expensive is not here the best. Cotton napkins are much to be preferred to linen ones. The linen allows the moisture to pass through and to saturate the clothing. The cotton absorbs and retains it. Of course the baby should not be allowed to remain wet after his condition is discovered; but even the brief time that must occasionally elapse before he receives proper attention makes the fact that wet linen is much colder than wet cotton of no small importance. Put the extra money that linen napkins would cost into a larger number of cotton ones. It is almost impossible to prepare too many.

Habit is a great helper in keeping the baby dry, just as it proves to be in making him sleep and eat. If for a few weeks an infant is changed promptly every time he requires it, he will learn to grunt and fuss significantly whenever he is wet, conveying as clearly as if by word that he wants and expects to be relieved from discomfort.

Among the habits which materially contribute to the maintenance of good health, and which should be early established, is that of regular movements of the bowels. It is possible to accustom even very young babies to using the chair—some experienced nurses maintain at as early an age as two or three months. But even if it is possible, it is probably undesirable for any baby under six months of age either to sit on the chair or to be held in an upright position over it. It is possible that less harm results from the usual method of using the napkin than from this constant disturbing and handling. But somewhere between the ages of six months and a year, according to the strength and physical development of the individual, this habit of cleanliness may usually be established without injury to the child and with less difficulty than at any later age.

The mother or nurse should, however, take the whole burden of the lesson upon herself, and not lay any part of it upon the baby, for whose feeble brain the task of remembering or indicating the necessity for the chair is an unwise strain. The whole aim of the first six months should be to make the body grow and to keep the brain quiet.

A child less than a year old must never be disciplined in the sense of being expected to make a conscious mental effort. It should be trained only so far as habit, and not conscious effort, aids us, and then only in those physical functions, such as sleeping and eating, which are with the child purely animal. That is, we must not at this immature age keep the child on the chair for any length of time, or endeavor to impress upon his mind the necessity of emptying bowels or bladder at that particular moment. We should rather shoulder the responsibility ourselves, and so carefully time our efforts that they coincide with the natural inclinations of the baby, thus making the lesson physical rather than mental.

All disciplinary efforts for the first two years of life should be in the line of establishing and strengthening physical habits. At the same time we must make every effort not only not to encourage, but actually to retard, any complicated mental effort. It would be better to delay the formation of the habit of using the chair until the end of the first or the beginning of the second year than to impose on the child any sense of responsibility, or to encourage it to any conscious effort to communicate its wants. If we can so accurately anticipate the child's wants as to hold him over the chair for a minute just at the right time, and have the inspiration to continue this practice with judicious regularity, then the physical comfort alone will in most cases induce the child to respond to one's efforts.

CHAPTER VI

DIGESTIVE DISORDERS

For any obstinate form of irregularity of the bowels medical advice must be sought. Of the two extremes, diarrhœa is from the medical standpoint the more serious, and demands more immediate attention. But for a thoughtful mother, concerned for the ultimate welfare of her child, both are equally significant, and convey wider lessons than are usually mentioned in connection with either. While the medical treatment of these troubles in their various manifestations must needs be left to the physician, the discussion of their ultimate consequences occupies an important place in any scheme for the prevention of disease.

Diarrhœa, in nearly all its forms, from the

most simple to the most serious, is now believed to indicate an excess of noxious microbes somewhere in the intestinal tract, and points directly to grave errors in the method of feeding. It also demands prompt treatment by which these microbes may be washed from the digestive organs. The mother should make it her most imperative duty to accept the first warning of this kind, and to rectify mistakes in feeding before repeated attacks of the same nature impair the integrity of the mucous membrane of the part upon which the microbes have fastened. It is a matter of perhaps twenty-four hours to clear the system of microbes at the first attack, while it may require months or years to restore the organ to its normal condition after it has once been injured by the repeated attacks of these germs. To nourish the child while this reparatory process is going on is a problem too often beyond the power of maternal affection or medical skill to affect.

Constipation teaches an entirely different

but hardly less important lesson. It usually indicates that the child receives too little nourishment, or nourishment which is too concentrated. It may sometimes in a degree be temporarily even a favorable sign. A child whose diet is changed from food which contains injurious germs to good germless food will usually become constipated for the very lack of this dangerous irritation which the microbes produced upon the intestines. Any sudden impetus to more rapid growth, whereby the body assimilates a larger proportion of the food taken, may also be the cause of constipation. It needs careful study and discrimination on the part of the mother to ascertain whether the constipation is chronic and deleterious, or whether it results from conditions which are temporary only, and will in the end even prove favorable. If the constipation is chronic, and shows evidence of increasing, a change of food is as important to the well-being of the child as it was in the case of diarrhœa. If it is merely temporary, patience and the application of simple remedies will soon correct the trouble, and demonstrate conclusively that this was simply the precursor of an improved condition. In case of constipation, from whatever cause, it is wise to increase to the point of toleration the daily amount of water that the child consumes.

If a child is over a year old it should be taught to drink from a cup, but even a young babe should be given water from a spoon or bottle several times a day. In addition to this, a movement of the bowels should be artificially induced, either by glycerine suppositories or oil-and-water injections at a regular hour each day. This responsibility should rest on some one person, either the mother or a trustworthy nurse, and must never be omitted or varied if the constipation is to be overcome. Among the physical aids for the cure of constipation are vigorous out-of-door exercise for older children and abdominal massage for babies.

The habit of drinking daily a quantity of water is one that is valuable in many ways. Its importance is seldom sufficiently emphasized. It is not enough that the child should take an occasional glass of water, or that the babe should be given a spoonful as a rarity. But the *habit* of water-drinking is essential to the well-being of every child. Most children will occasionally ask for water at meals, or will take a swallow of ice-water when they see others drinking, or will enjoy water with lemon, or fruit, or jelly, or sugar, or flavored with tea or coffee; but water pure and simple it seldom occurs to a child to demand, or to a mother to offer, although of all foods this one is the most important, and no other contributes so directly to the health and growth of the child. The tiniest baby should be given a teaspoonful of water many times during the day; and if at night it takes water from a nursing-bottle, it will require during several hours no other nourishment. A child two years old may with advantage drink at least a pint of water every twenty-four hours, and a child from three to four years old will not infrequently consume a quart of water in the same time.

All water fed to a child should have been boiled, and must be kept in a bottle or carafe that can be closely stoppered. It should neither be warmed nor cooled, but should be given to the child at ordinary temperature as it stands in the living-room. It should always stand within sight of the infant, and within reach of an older child. Where it is necessary to go down-stairs or into another part of the house in order to obtain a drink for the child, it usually has no drink at all except at such times as its thirst becomes intense. It is not necessary or advisable to give water to a child during meals, but at other times it may safely be allowed to drink as often and as much as it will. It may even be encouraged to increase the amount, if the water that is used has first been boiled and is of the proper temperature. We cannot of course force a child to drink, nor is it pleasant to over-urge such a necessary operation. But by having water always at hand we may make drinking easy, and by providing a pretty cup, or making some merry play, we can go farther and make the drinking of plain water really attractive until the habit is firmly fixed, when it will regulate itself.

CHAPTER VII

SENSE-DEVELOPMENT

THERE is a conviction prevalent that a child which is left so completely to itself as the method previously described would indicate will, at least during babyhood, be slow and dull, and will feel bored for lack of interests. But in reality the undisturbed child, while serene and sweet, is in every waking moment also unusually and uniformly active and gay. He discovers and interprets gradually and naturally both the small ego and the great non-ego; and since he discovers them from the standpoint of his infant observation, not forced prematurely from the point of view of the adult mind, he will find in the process endless amusement without disturbance or excitement.

The sense of touch is the last of the human

powers to be wiped out by the on-coming of death; it is also the first to develop in the new-born infant. The first sensations of this outer life are usually not agreeable to the new-born child. His feeble wail, a protest against the wide unknown, seems to invite our compassion, and usually tempts the attendants to offer injudicious petting. If, however, the first feeble sense of touch is used to give the child a point of contact with the new world, a baby even a few hours old, unless it is in pain, will be comforted if it is allowed to clutch in its tiny fist the finger of some friendly hand.

The prehensile powers of a baby are proportionately much greater than those of a mature man. Many children, when only a few weeks old, are able to sustain their weight by hanging by the arms. Through this ability to grasp and hold whatever comes in contact with their curving fingers comes their first self-taught lesson, and their first means of diversion and investigation.

To the first sense of touch the average child a week old adds a feeble consciousness of the sense of sight, and begins to follow moving objects with his eyes, or to observe anything that is shining or bright. At the age of a month it will turn its head and follow moving sounds. When six weeks old a child will begin to distinguish, not by sight, but probably by touch or smell, its attendants one from another. When ten weeks old it will so far in its feeble brain have formulated the fact of friendly attention that, if well cared for, it will no longer cry or wail unmeaningly or indiscriminately whenever it feels hunger, pain, or discomfort, but will grunt and scold, with cheerful and evident confidence that its wants will be considered as soon as made known. Some children at this age will fasten the eyes upon the person who usually attends to such necessities, cooing and chattering in a seeming effort to convey by the inarticulate language of infancy their personal wants.

When it was eleven weeks old, one of these

judiciously neglected babies was heard to laugh out so loud as to frighten himself; and another at the same age was proved to notice, distinguish, and show preference in colors, indicating great pleasure in dull blue, and distress and physical discomfort at bright pink. At three months of age one child passed infantile judgment upon musical tones, screaming with apparent rage whenever the sharp tones of a hand-organ rose from the street, but cooing and laughing with delight whenever a fine piano in an adjoining room was touched. Before this age the average child has also discovered himself. First he finds his hands, and they afford him many a day's amusement and furnish valuable lessons in natural history. He discovers successively that they move, that they belong to him, and finally, more wonderful still, that they move at his own volition. These movements are at first aimless and without purpose, but the gradual effort to convert them into intentional motions entertains many a baby for days or weeks. The discovery of the head furnishes perhaps the greatest wonder and amusement to the child, since, unaided by the sense of sight, he must explore that region with the help of the half - trained hands alone. As the little fingers wander round and round the tiny dome, a look of interest and comprehension will gradually replace that of astonishment, and this transition marks another distinct epoch in the natural mental development of the child.

When we entertain and amuse an infant we do not help in its essential development, but rather hinder its normal growth. We excite and weaken it; but nature teaches and strengthens the infant mind. A child three months old, already observing a difference in sounds and in colors, and formulating, even though feebly, the personality of those around him, faces literally the whole world of material sensations, and will gain more new information by his own unaided perception than either the father or mother could possibly acquire in a much longer time without an attack of nervous exhaustion. We cannot prevent this natural, rapid development, nor would we wish to do so; but we need to avoid with the utmost care either interfering with or accelerating its progress. All the environment of a child should remain as nearly as possible the same day by day. New rooms, strange faces, unusual sounds or sights, should be avoided in order that he may learn to know the "I" and "not I" in their simplest forms, and with the minimum strain upon the nervous system.

When about a year old a child enters into its first comprehension of the power and value of language, which is the door of intellectual life. With this acquisition it leaves babyhood behind and crosses the threshold into childhood.

CHAPTER VIII

RATIONAL FEEDING

Of the many elements of caution that during the first year of life contribute to the good health of the child, care in feeding is the most important; perhaps more important than all the other elements combined.

In considering, as we shall, one kind of food only, we do not desire to ignore the fact that there are other foods which have in many cases proved valuable, and upon which children have been successfully reared. Sterilized milk is, however, now recognized as the *best* artificial food for children, and where we can obtain the best it is manifestly unwise either to consider or to use an inferior article, even though it may have intrinsic worth.

The general insufficiency of the breast-milk of the mothers of the present generation, and the tremendous drain that lactation makes upon the average woman, put the natural food of infants in many cases out of the question. It is universally acknowledged that good breast-milk is superior to any other food for an infant; but we must at the same time recognize that not one mother in ten can or ought to nurse her child for more than the first few weeks of its life. Therefore artificial food must be discussed not as a mere substitute, but as the general rather than the exceptional food for infants. Before the discovery of sterilized milk many a mother, at the risk of her own health, and in spite of the fact that her milk was insufficient for the demands of the child, still persisted in nursing it, rather than incur the perplexities and dangers of the old system of bottle-feeding. Now, however, with the use of sterilized milk, there is no longer danger of any sort, and perplexities may, by an exact system of feeding, be reduced to a minimum.

Good breast-milk agrees with all children, and all except those essentially diseased thrive upon it. Good sterlized milk, properly prepared, proves its similarity to mother's milk in nothing so much as in the fact that it, too, agrees with all children, of whatever age or condition, unless they are so acutely ill that all milk must for a time be abandoned. In such rare cases a change to water, barley-water, gum-arabic water, or beef-juice for a time long enough to clear the intestinal tract of the collection of microbes is all that is necessary; twenty-four to forty-eight hours are usually sufficient. The feeding of the sterilized milk may then be gradually resumed by using in the barley-water a very small quantity of the milk, increasing it by a spoonful at each feeding, until the average proportion of milk and water is again attained.

Great pains have been taken to collect statistics on this particular point, and in every instance where good, rich, thoroughly sterilized milk has been reported persistently to disagree with a child, some grave defect in the manner of administering it has been detected. In one case laid down as an argument against sterilized milk, the baby was fed while lying flat on its back, through a nipple in which the hole was so large as to allow a rapid stream of milk to flow down the child's throat and almost to strangle it. Naturally milk taken with such rapidity violently disagreed with the child. In other cases the milk was found to be administered quite cold, or too hot, or not properly diluted; or it was taken from a bottle that had been too long opened; or it was fed through a flexible rubber tube whose interior uncleaned surface polluted the milk as it passed. Many children just taken from a breast whose milk is thin and unnutritious crave and will rapidly swallow an equal quantity from the bottle; whereas this milk, which is many times richer than the former food from the breast, should of course be given in smaller quantities. If given in equal bulk it will inevitably make a child ill. It is not, however, sterilized milk that should in such cases be condemned, but the breast-milk, whose insufficient richness taught the child to demand a quantity sufficient to produce chronic dilation of the stomach, and to create an abnormal appetite. Having made notes of thousands of cases, I have yet to find one child who would not thrive on sterilized milk if properly administered.

There is, therefore, no need to discuss other foods, since what is acknowledged to be the best is always available, unless the child is so ill that it can digest no food at all. The advantage of breast-milk over even sterilized milk arises from the fact that there is less room for error in its administration. Nature prepares the food of the breast, and Nature teaches the child the method of obtaining it. She leaves no room between the breast and the mouth for mistakes of any sort. Good sterilized milk is as germless and safe as good breast-milk, but in its administration there is room for stupidity and carelessness that may neutralize its good qualities. Each detail in the preparation and administration of the bottle is of infinite importance. And, reasoning backward, the mother may be absolutely certain that if the effects are not good the preparation is defective.

Some children do not thrive even upon the best food, whether it be artificial or from the breast. In ninety out of one hundred such cases the cause lies in the system of general management, which has resulted in an overstimulated nervous system. A baby is, and should be, solely an incorporate stomach. The digestive processes should be the main object of its existence, and nothing else should interfere with this operation, or detract from the strength put into it A child less than a year old, if it is over-excited and over-entertained, continually diverted and amused, cannot properly digest its food. And when once it is thus over-stimulated, the artificial craving for excitement is established as a habit, and the child demands or seems to require a continuation of the activity which is undermining its physical strength and impairing its digestion. In such a condition it is, of course, of no use to change the food, or to hope to find anything that will nourish the child. It is even wiser to reduce rather than increase the quantity given, since the amount of undigested food in the intestinal tract determines the degree of danger. If it is not too late for any remedy, the only one that can possibly contribute to the return of digestive power is one that will induce less nervous activity. The child should be kept in a quiet, darkened room, under the care of one person, the mother if possible, and the best medical advice should at once be procured.

The dangerous nervous disturbance with babies comes from too much handling, overexcitement caused by a child's seeing too many strangers, and the too early stimulation of the organs of sight and hearing; little or irregular sleep; and too many, too early, or too noisy out-of-door excursions. It is a condition more easy to avoid than to cure. Its most serious result is the shattered nervous system, which the children who survive these attacks carry throughout life.

Some mothers, having by lack of care in artificial feeding, or by over-excitement of the nervous system, impaired the child's natural digestive powers, hope to avoid the consequences of their own carelessness by the employment of a wet-nurse. This practice is, however, becoming more infrequent, and, it is to be hoped, will soon be obsolete. With the present improved and entirely safe methods of artificial feeding, there is no case where its dangers are not less than those incurred by the employment of the average wet-nurse. An ideal wet-nurse may exist, but is never to be found when an emergency demands. It is always difficult, and in large cities impossible, to trace the antecedents of the women who apply for such positions. There is a strong chance that they are diseased, almost a certainty that they are immoral, and no hope that they will give the child any judicious or systematic training. Any one who has had experience with this class knows only too well the impossibility of restraining them in drink or diet, and no one can be certain that they will not dose the child to secure a night's sleep for themselves. The natural and inevitable risks that a baby encounters in its first years of life are multiplied many times whenever a wet-nurse is employed, and it is an unusual combination of circumstances that can justify a mother in incurring such unnecessary dangers.

CHAPTER IX

STERILIZED MILK

ONE-FOURTH of all the deaths in the United States are of children under one year of age; and nearly one-half, in round numbers 400,000, are of children under five. In cities this proportion rises during the warmer part of the year, until one-half of all the deaths are of babies less than twelve months old. The majority of these children die of diseases caused by germs introduced into the system in the uncooked milk and water, which constitute the sole diet of many infants, and the principal food of all young children. Intestinal diseases, counted non-contagious, carry off by far the greatest number. Experience has proved that these troubles may be modified, or in many cases entirely eliminated, by the use of germless food. By feeding the child only milk that has been sterilized, and water that has been boiled, we cease to feed the disease and begin to nourish the child.

Sterilized milk is comparatively a new discovery, and the difference between its use and abuse is not yet distinctly defined in the public mind. The apparent simplicity of its production has misled many physicians, as well as mothers, into applying the name to an article which possesses none of the virtues of sterilized milk.

American investigation on the subject has been extremely crude, and so far is still totally inadequate as a basis for sound conclusion. Fortunately, in Europe the subject has received due consideration. German scientists especially have given much time to the investigation of the effects of various kinds of milk in the intestinal diseases of children. Also Tyndall, Lister, and Pasteur have carefully studied milk in all its natural phases of composition and decomposition. They have gone to the very foundation, having themselves taken the milk from the cow, under varying degrees of atmospheric impurity, and carefully noted in each case the favorable or unfavorable environment; and they unanimously declare that all milk from a healthy cow is absolutely pure and germless as it flows from the udder, but that its composition, its animal heat, and its exposed surface, all combine to render it a most favorable medium for the cultivation of bacteria. On the other hand, the atmosphere of the ordinary stable, swarming as it is with germ life, at once furnishes in plentiful measure the microbes, which, coming in contact with the milk, instantly begin to multiply at an appalling rate. In any common stable milk cannot remain free from infection even while it is flowing from the udder to the pail.

Koch, Escherich, and their celebrated co-workers have supplemented the investigation of milk in its natural condition by valuable studies of the germ life which is found in the intestinal tract of an infant, and have noted its variation in health and disease. They conclusively demonstrate the poisonous effects of impure and germ-laden milk upon the delicate digestive organs of a child.

All these scientists conclude that there is no strictly pure milk except that taken directly from the udder of the cow; that the milk delivered in cities, whether twelve, twenty-four, or thirty-six hours old, is swarming with microbes; and that it varies only in the degree of its dangerous properties. Therefore they declare that all milk fed to children and invalids should first be carefully sterilized in order to destroy its countless bacteria, which otherwise would be introduced directly into the system.

In Germany the danger of using unheated milk is so clearly comprehended that legal enactions regarding it are becoming every year more stringent, and it is already difficult for a traveller in that country to procure a glass of milk that has not first been steamed or boiled. In America the necessity of sterilization is not so generally recognized, nor, if we may judge from the reports in medical journals, have the results been so exceptionally good.

The inconsiderable proportions to which infant mortality has been reduced in the public institutions where even partially sterilized milk is used indicate, however, what blessed results might be hoped for if the milk supply was controlled by judicious legislation, and the quality and condition as it is delivered to consumers regulated by law.

Contrary to the more mature opinion of European authorities, an American physician will occasionally affirm that sterilizing milk renders it less digestible, because it coagulates the albumen. Cooking meat and eggs coagulates the albumen, but we do not therefore conclude that meat and eggs should be eaten raw. On the contrary, it is known that cooking meat renders it more digestible, provided always that it is not overdone. So in like manner the digestibility of sterilized milk depends upon the degree and duration of the heat which is applied. Milk that is swarming with microbes cannot be sterilized without prolonged heat applied on successive days. But fresh milk can be freed from germs with such a moderate application of steam, that when once the milk is re-aerated it is difficult to distinguish it from the new milk of the milking-pail.

Sterilized milk has usually been recommended as especially valuable in diseases of the stomach and bowels. Its highest value is not, however, as a medicine, but rather as a food. Favorable as are the results of its use for sick babies, its best work is always with children of average health and heredity. Its chief value is not in the cure, but in the prevention of infantile disorders. These, as a rule, attack those only whose vital powers have through some form of malnutrition been reduced below par.

The artificial foods that preceded sterilized milk in popular favor were all defective in one or the other of two ways—they were either unsafe or unnutritious. Those of the first class, comprising nearly all sorts of milk diet, furnished the proper and natural elements of nutrition, but were dangerous because they contained such abundant germ life that the child who took them was seldom well and often violently ill. Those of the second class included the patent baby-foods and condensed milk; they eliminated the elements of danger arising from bacterial infection, but failed to furnish sufficient nourishment to meet the demands of a growing child. Each class avoided the danger of the other, only to incur as great a danger peculiar to itself.

Milk as a food furnishes all the elements necessary to life and growth. Now that it also can be made free from germs, it is, when properly prepared, an ideal food, and its discovery has revolutionized the whole system of infant dietary. It is above all others the food which appeals to the common-sense of mothers. It is not artificial or mysterious in its composition. Sterilization is merely a method of restoring milk to its natural germless condition, and retaining as far as possible its normal elements of nutrition. The process is simple, and the tests of its efficacy are easily applied without scientific training.

The common belief that the primary object of sterilizing milk is to prevent it from souring is misleading. Milk that is in danger of becoming acid before it can be used is already unfit to feed to infants. The important object to be obtained by sterilizing is to destroy as soon and as thoroughly as possible the bacteria, which otherwise continue to feed upon the milk and to destroy the fat globules, the constituents containing the elements essential for the nourishment of the babe. Most of the milk used for children is, even when fresh, deficient in fats, and the uninterrupted action of the germs renders it simply starvation rations for any growing creature.

Any mother can test the sterilized milk she uses and discover if it fulfils the two requirements of an ideal food for infants. Without the aid of chemist or microscopist, she can determine if it contains adequate nourishment and is free from germs. By pouring a small quantity of the milk into a graduated test-tube, and setting it aside for twenty-four hours, she may learn just how much cream it will yield; and by placing one of the bottles in the temperature of a living-room for two or three days, she can ascertain if the milk is sufficiently well sterilized.

Most children are fed too much in bulk. The milk they drink is not rich enough to satisfy, with any normal quantity, their healthy appetite. To approximate to good breast-milk, we must start with cow's milk that will yield one-fourth its own bulk in cream; this, when diluted with an equal amount of water, will yield a food that is safe, nourishing, and entirely adequate to all the demands of a hungry stomach. Fed on ordinary city milk, many children slowly die of starvation, or become in time the victims of chronic illness resulting from malnutrition. An infant may be fed to repletion and yet be poorly nourished. Scores of even breast-fed babies are half-starved without ever having suffered from hunger. Malnutrition is indicated by late dentition, poor bone formation, a tendency to rickets, broken sleep by night, general fretfulness by day, a susceptibility to colds, and a liability to catch all the prevailing diseases in consequence of lowered vitality.

Immunity from disease is especially important during the first year of life, since a child's power of resistance is then at the lowest ebb, and its susceptibility to infection at the maximum. Statistics prove that with every month of existence a child's hold on life is strengthened. Four times as many children die in the first as in the second year of life. Good health means not present blessing only for a baby, but every day's exemption from disease is so much increase in the surplus vital energy that shall render the child capable of resisting infection in future. And food is the agency by which we must build up a strong foundation of permanent good health.

The fact that one cannot produce perfectly sterilized milk at home is not an argument against its domestic preparation, but is in reality the strongest of all pleas for a careful steaming of all the milk that is to be used in the family. If the germs are so difficult to destroy, so active and prolific, then the greater is the necessity for killing as many as possible before introducing them into the digestive system of man or child.

Many mothers incur extreme and unnecessary risk from the belief that when the milk is steamed it is thoroughly sterile; whereas if they realize that it is only partially sterilized, that the germs only, and not the spores, or seeds, are destroyed, they would exercise greater caution in its care and administration, and hence take fewer risks.

With this qualification in mind, we may with clear conscience proceed to discuss the best methods of home sterilization. First, the age and quality of the milk must receive careful consideration. During every hour in which the milk remains exposed to the atmosphere, or is shaken by the motion of transportation, it deteriorates, and the bacteria, which find lodgment in the milk almost as soon as it leaves the cow's udder, multiply in geometrical ratio. The common hay bacillus, found in all stables, and consequently in all milk, multiplies so rapidly that at the end of twenty-four hours its descendants number 10,000,000,000. These germs live upon the milk, and the microscope demonstrates that under their operation the fat globules composing the cream gradually disappear, few or none remaining after the fourth day. With sterilized milk, on the other hand, no change is visible, even with the microscope, except a tendency of the fat globules to coalesce, a process popularly known as condensation of the cream. Therefore in fresh milk we find few microbes and many fat globules; in old milk, many microbes and few fat globules.

Cow's milk differs from mother's milk in that it contains more casein, or cheesy matter, and less of the necessary fat. To restore the natural proportion we need to use milk rich in cream, as from the Jersey or Guernsey cattle.

The process of sterilizing milk is simple in detail and easy to describe. The burden of the work lies in the effort to maintain uniform and absolute cleanliness throughout the whole process. Not only must visible dirt be abolished, but the cleanliness of every article that is to be used must, even to the searching eye of the microscope, be unimpeachable. We need first to discard any apparatus that is complicated in structure, or has parts inaccessible to air and light; and any instrument that might furnish a favorable nidus for the propagation of germs should be at once rejected.

All bottles to be used either for sterilization or nursing should be spherical in shape. Sharp corners in the interior of a bottle are difficult, if not impossible, to clean, and may at any time retain an invisible particle of milk, to become the focus of tyrotoxicom poison, which cannot but prove fatal to a child. Only short nipples that are easily inverted are allowable. There is no virtue in sterilized milk if it must, in its passage to the child's mouth, flow through a long rubber tube lined with colonies of germs. No sponges or brushes should ever be employed for cleaning the bottles, for after they are used they themselves furnish more germs than all our cleaning can remove. Every bottle emptied of milk should be rinsed in cold water, and then submerged in a pail of water in which has been dissolved an ounce of baking-soda. When the day's collection of bottles is to be thoroughly washed, preparatory for refilling, it facilitates the process to have ready at hand a pail containing white castile soap dissolved in water, to which has been added a tablespoonful of ammonia. With this one may use a clean bit of cloth, tied to the end of a wire or stick; or may shake in the bottle a piece of raw potato, small pebbles, sand, or rice grains. Cloth, potato, pebbles, or rice should, however, not be used a second time. Whatever is employed must be renewed each day. After washing the bottles should be rinsed with boiled water, and then immediately filled with the milk to be sterilized.

The principle of sterilizing is simply to keep the bottles of milk in boiling water or live steam for long enough time to kill the germs. This may be accomplished with an ordinary tin boiler and steamer used for cooking, but is more conveniently done with some one of the numerous sterilizers which are offered for sale in large towns. Of these the best known are the Soxlet Sterilizer, in which the bottles are partially submerged in boiling water, and the Arnold Sterilizer, in which they are inclosed in a chamber of live steam. Both these sterilizers now come furnished with round-bottomed bottles, which are not only more easily cleaned, but are less readily broken by the repeated heatings, than the flat bottles.

A variety of stoppers have been successively used—rubber, cork, and cotton; but for home use nothing equals for convenience and efficacy the double Soxlet cork of rubber and glass. The initial expense is greater, but the saving of time and the superiority of result more than compensate for this expenditure.

The length of time necessary to sterilize milk depends upon its age, and varies with the apparatus that is used. The time, as given by various experimenters, runs from thirty minutes to three hours. It is wise in the beginning of the work for a mother to set aside one bottle every day to test the efficiency of the process. The test bottle should be placed in a room whose temperature is from 40° to 70° Fahrenheit. If the milk turns within forty-eight hours, the steaming is insufficient. If the milk remains good for from two to three days, it is safe to feed to the child. Milk sterilized at home, if it will keep longer than this, has usually been over-heated, and thereby so much changed in composition as to lose some of its value as food.

In diluting sterilized milk, one should always use water that has been boiled; for ordinary drinking-water is one of the most favorable elements for the propagation of bacteria, and may any time add again to the milk just those germs which we have been at such pains to eliminate.

CHAPTER X

FROM INFANCY TO CHILDHOOD

AFTER a child is a year old the measures to be adopted for the prevention of disease and the preservation of uniformly good health can no longer be given in such simple and universal rules. A young infant is an unreasoning animal, and with it the physical conditions alone need to be considered. Its food is simple and simply administered; and beyond the general desire for physical comfort and satisfaction, it expresses no preferences and conveys no criticism of our methods.

But after it is a year old a child begins the differentiation towards a more complicated existence. After that age a child is no longer simply an animated stomach. It has already found its hands, and learned that they can clutch and grasp; it has discovered its feet, and is fast learning the art of locomotion; it has become an apt pupil in the lesson of language, that instrument of all intellectual progress. It has formulated the ego; and after the knowledge that "I am" is once defined it soon conceives the second lesson of "I want." Within a short time the "I want" is followed by an "I ought," and with this last conception the triple development of the physical, mental, and moral natures progresses. Nor in any consideration of childhood, from whatever standpoint, can these three simultaneous and interdependent lines of development be separately considered. If we discuss intellectual education, we find its success ever dependent upon the physical condition, and incapable of the highest attainment except in the presence of a normal moral sense. If we consider moral development, we find it inextricably complicated with that of the intellectual and physical natures.

So in considering, as we at present aim to do, the measures that must be taken during childhood for the preservation of the best health and the practical elimination of infantile diseases, we find it impossible to consider the physical alone, but, even at the risk of seeming superficial, must touch, at least in many points, upon the mental and moral training of the child. Its physical health is always dependent upon proper mental and moral training. Every physician, for instance, encounters in his practice among children cases of illness which terminate fatally simply because the child is so wilful and undisciplined that his struggles against the prescribed and necessary course of treatment turn to the fatal issue the evenly balanced scales in which are weighed the alternatives of life and death. So the chronic habit of disobedience or deceit on the part of the child may neutralize the parents' best efforts for its physical improvement. And fretfulness, generally a result of disease, is not infrequently, when it becomes a fixed habit, also one of the causes of illness, or at least of chronic ill heath. Any discussion, therefore, which deals solely with the physical precautions for the prevention of disease must be absolutely inadequate. To obtain the desired result it is necessary to touch upon mental education and moral training, at least as far as they are involved in home discipline and home amusements.

It is also important that parents who would comprehend and enforce the necessary measures for the preservation of their children's health should be familiar with the standard scientific authorities, which form the basis for any valuable educational discussion. Every mother who aims intelligently to train her child should be familiar with those works of Spencer, Preyer, Perez, and Fröebel which treat of child nature and child needs. Without some such preliminary reading, it is difficult for a mother intelligently to follow any rules that may be laid down. Every child must, in many points, prove itself an exception to the general rule by failing to conform to the average standard; and in order to appreciate to what degree this divergency is vital, and in what sense it is unimportant, one needs to comprehend what the average standard really is, and to be familiar with the scientific laws underlying any special rules for education. If a more perfect knowledge is desired, and if the parent would be competent to make rather than to follow rules, to go back to the first principles underlying all development either of individual or of race, this knowledge can be obtained in no way so well as by a general study of the fundamental theory of evolution.

It is well understood among scientists, and now generally accepted by all intelligent people, that a child closely approximates, in many of its attributes, to the lower animals. Children are neither angels spoiled in the making nor are they to be counted as illustrations of natural depravity. They are at first simply animals of a lower order in the scale of development, in whom the mental and moral qualities are nascent, and of whose present needs and future possibilities we can obtain no adequate conception except by an intelligent study of the lower species which they resemble. Each individual child follows step by step, in its personal growth, the path by which the race has progressed to its higher destiny. It begins life, prenatally, as an aquatic animal. Its first attempts at locomotion are, like those of its brute ancestors, made on all-fours, while it possesses naturally, during the first year of life, prehensile powers greater than it can ever afterwards attain without the training of an athlete, and equalled only by those of its cousin, the ape.

The value of a knowledge of evolution in its relation to the education of a child is too many sided to receive here more than a passing comment. But as it is, in at least one of its phases, the foundation upon which we must build our educational work, that phase, if no other, should command our careful consideration; for the history of evolution alone can indicate which traits of a child's nature are permanently and increasingly dangerous, and which are only temporarily disagreeable. It is of vital importance that we should withhold our discipline, and become, if possible, blind and deaf to those natural, transient, and universal faults of childhood which, the theory of development indicates, will cure themselves. Of no less importance is it that we reserve our attention and influence for those errors which grow with maturity, and that we stamp out with unremitting energy any serious and permanently evil habits in their very inception. But until the parent is able himself to distinguish a fault from a sin, a natural, healthy impulse from a depraved tendency, it is not possible that he can give any vital assistance to the child he aspires to train.

To illustrate: All children are noisy. Noise is the natural expression of natural animal vigor, and is necessary to the healthful development of all young creatures. Therefore, while a noisy child must occasionally be restrained, it should never be punished, nor should mere harmless noise be made to seem to the child a thing to demand reproof. Moreover, we should even encourage this natural, healthful tendency by providing a time and place in which its indulgence may be unrestrained. Unfortunately, to most children noise is made to appear the unpardonable sin, than which no error of the most serious moral nature is more constantly reproved.

Even those childish faults which seem to the mature mind to involve a serious moral question are seen, when judged by the comparative standard of evolution, to be likewise temporary and unimportant. For instance, untruth does not, in a little child, usually indicate any moral obliquity. It either arises from the purely animal instinct of concealment, or it is a result of inaccuracy of observation or the outgrowth of an over-vivid imagination. From whichever of these causes it orignates, untruth has a tendency to cure itself with the development of the intellectual powers, unless the direct heredity bias towards it is exceptionally strong. In any case, punishment for falsehood is of doubtful wisdom. It often gives a child who was before simply imaginative its first clear-cut idea of what falsehood really is, and leads a child naturally deceptive to cultivate more subtle forms of untruth. It is quite possible, and usually easy, to teach a child absolute truthfulness by leading it kindly and gradually to distinguish between reality and the creations of its own imagination. If, in addition to this, the young find unvarying truthfulness in the older people about them, if they understand that their parents regard as sacred every promise made to them, then their animal imitativeness will be the strongest aid in cultivating the same quality of truth in them.

Cruelty, selfishness, destructiveness, and violent physical manifestations of bad temper are among the evil tendencies which are strongest in childhood, and which have a natural tendency to correct themselves with increasing maturity.

The majority of the faults of childhood, indeed, result from this predominance of the simple animal instincts; and if judiciously ignored or mildly corrected, will drop away, to be replaced by more desirable qualities, as quietly and inevitably as the petals of the fruit-blossom drift to earth when the heart of the flower forms the young fruit.

CHAPTER XI

NORMAL OBLIQUITIES

Out of consideration for the peace and comfort of the remainder of the family; it is often necessary to correct and sometimes to chastise the child for undue indulgence in even natural traits. But punishment should be tempered by the comprehension that savage instincts are, to a greater or less degree, normal in all children, and that they will of themselves constantly diminish in strength.

It is also important to remember that the attributes which are most disagreeable in childhood are really the most valuable in maturity. The noisy, incessant activity of the child develops into the energy of the man. Destructiveness in the young is the elementary manifestation of the investigator's spirit. Troublesome obstinacy grows into perseverance; and over-strong will-power, which often thwarts the parents' best efforts for the child's discipline, becomes, when properly trained, a most desirable quality in maturer years. Intentionally to diminish a child's power of resistance, his persistence, or the force of his will-power, is deliberately to rob him of the best capital he can ever possess. We may and must judiciously limit the exercise of these powers in order to make life with a strong-willed child endurable; but the discipline is of temporary value only, and should be counted merely a convenience for ourselves. It can have no permanently valuable effect upon the child's future, except in so far as we convince his intellect, and demonstrate to his satisfaction the value of self-discipline.

Among the traits which are not natural to childhood, which will, unless promptly eliminated, increase rather than diminish with advancing maturity, are those arising from a too intense self-consciousness—a trait which is foreign to healthy animal existence, and is one of the penalties we pay for our civilization. Among its manifestations we may note that of excessive bashfulness, or its counterpart, offensive boldness; an eagerness for attention and commendation; the demand for continual amusement or over-excitement; and a chronic discontent or persistent fretfulness. All these traits indicate an over-stimulation of the nervous system, and a precocious concentration of the childish mind upon itself. While disagreeable in the child, they are increasingly offensive in the man, and are always antagonistic to the highest personal development. They should, therefore, at their first manifestation, receive the parents' most careful attention, and no pains should be spared to prevent their continued growth.

There are also certain demonstrations of temper which are decidedly and increasingly dangerous. The sudden, short-lived temper of the baby, with its natural animal protest against that which is displeasing, is, as we have said, transient and in no aspect serious. But the temper which manifests itself later, which is brooding, or sullen, or malicious and lasting, should at its very first demonstration be counted as a danger signal, indicating the necessity for immediate and constant repression.

A strong analogy exists between the growth of the body and the development of the mental and moral natures. If the body is properly fed, it will seldom indicate any condition of disease; and if a child is provided with healthy, rational occupation, abnormal conditions of mind and heart will rarely be indicated. We are only now learning the elements of the science of dietetics for children, and outside the kindergarten little or no attention has been paid to the occupations of children, which should in like degree keep the higher side of their natures in normal condition. A better understanding of the proper occupations which would help to maintain a child in a condition of sound mental and physical health must come through a better knowledge of what has been counted as abstract science. Only by the comparative methods of evolution can we understand and guide the development of a child's nature.

Evolution, as it teaches of heredity and environment, of growth and decay, of individual and race progress, must, as we have indicated, guide the coming generation of parents into that wider, wiser knowledge which is the foundation of all effort towards a more rational system of education. Each parent has his separate and individual work to accomplish in studying the heredity of his own children, and in anticipating and preventing the manifestations of recognized ancestral weaknesses.

Variation is the universally recognized

condition of all living creatures, human and brute. It has ever been one of the prime factors in the development of the race, and is as purely scientific and impersonal as Kepler's laws of the motion of the planets. In discussing heredity, therefore, the physician, recognizing the universality of the law of variation, simply seeks to ascertain what particular variation or combination of variations was peculiar to the immediate ancestors of the child under consideration, who is their natural and inevitable exponent. No child is the child of its father and mother alone. It is the grandchild of four ancestors, the great-grandchild of eight, and the great-great-grandchild of sixteen. It may revert to the individual idiosyncrasy of any one of these thirty ancestors, or even go further back and be most like some one of the multiplying numbers still more remotely removed. The responsibility for a child's deficiencies may not rest with either one of the two parents, and the remedy for these defects, which is in their hands, can only be found after careful consideration of the individual variations manifested in the ancestors.

We cannot prevent disease in childhood unless we know by this study of heredity what form of disease is most likely to attack the child, and therefore what part of the system requires special reinforcement. And before the best results can be attained, the word and the idea of heredity must become as essential and as inoffensive to the parent as they now seem to the physician.

The word heredity, for some inexplicable reason, is at present a proscribed one in almost every doctor's vocabulary. Physicians of the better class make a study of the individual heredity the basis of their diagnosis and prognosis in every case they are called upon to treat, but they often put the interrogations by which they obtain from the family of the patient the necessary data for a working theory without any direct mention of the word heredity, or any detailed explanation of its importance as bearing upon the case under treatment. Most of these men have learned from painful experience that the word heredity, to them of universal and therefore impersonal importance, conveys to the mind of the less scientific patient some ill-defined insinuation of physical taint or moral weakness. Until a clearer, cleaner idea of the importance of evolution and its bearing on heredity is popularly prevalent, the parent cannot become an intelligent second to the physician's effort to improve the condition of the child.

CHAPTER XII

VALUE OF MILK AS FOOD

As has been noted, the individual peculiarities of children, which begin to manifest themselves as early as the second year of life, act as obstructions to any set of fixed rules for the management of their daily life. Uniformity of management is usually possible with all healthy children under a year old, but after that period modifications must constantly be introduced, in order to shape these rules to the peculiar necessities of the child under consideration. As the child is no longer an uncomplicated being, the daily routine cannot be as simple or as rigidly adhered to as heretofore.

First, modifications in feeding become necessary, tending towards greater variety, in which the preference of the child becomes an element of success. A child's wishes must not, however, be allowed to play too prominent a part in the selection of food. Certain definite principles of diet must be adopted and adhered to as the foundation for healthy growth, and while the child's preference may to a degree modify them, it should not, as is too often the case, be permitted completely to overturn any rational system.

It is universally recognized as a fact that the majority of people, both children and adults, eat too much food and demand too great a variety. Milk is the natural food of all young animals, and should, with water, be the only article fed to the average child under a year old, and ought to be the principal diet of all children up to at least six years of age. The use of milk is sometimes too early abandoned under the plea that the child dislikes it. No young animal naturally dislikes milk. It is the normal and universal food of all young mammalia, and if it is repugnant to any particular child, it must be because in that child has been cultivated an appetite for less wholesome articles of food —usually for meat.

Milk might with advantage be used, to a much greater extent than is now usual, through all the period of rapid physical growth up to the time of maturity. Its importance as food for persons of nervous temperament of all ages is only now beginning to be understood. The difficulty of obtaining pure, fresh milk, at least in the centres of population, has limited its use and confirmed many in the conviction that milk does not agree with them—a mistake which has been fraught with serious results. The indulgence in an undue amount of meat, known to be injurious to the adult, is with the child absolutely fatal to any good results. The American idea that meat three times a day is necessary for the sustenance of life is positively disproved by facts, while the excessive restlessness and nervousness of the American people indicates clearly the penalty they pay for this error of judgment. Whole races of vigorous, healthy people have lived and worked and accomplished great things almost or entirely without meat, and in nearly every instance of the kind investigation proves that milk was the article substituted for meat.

We hear much of the oatmeal of the Scotchman. Dr. Johnson, indeed, defined oats as an article fed in Scotland to men and in England to horses, whereupon some one replied, "But where can you find such men or such horses?" But it is certain that an exclusive diet of oatmeal could never have induced such bone and sinew and brain as the Scotchmen boast of, unless it was taken always with a liberal allowance of milk.

In the Oriental countries where meat is proscribed by religious principles, rice or some other grain is commonly given as the staple food of the country. But it is necessary to remember that these people, like the Scotchmen, usually take their grain with milk. These Oriental people have attained not only good physical development, but have demonstrated such intellectual power and subtlety that in their eyes we of the Anglo-Saxon race appear as crude barbarians.

We need not go so far for illustrations of the virtue of milk as an article of diet. The medical profession of England and America recognize the easy digestibility, rapid assimilation, and non-exciting effects of milk by prescribing it almost universally as the sole diet in severe cases of fever or nervous prostration. In our medical journals are cited in detail, from the best authorities, instances of children who when upon a meat diet displayed violent or vicious tendencies, but when changed to a diet of milk passed rapidly into such a gentle, non-irritable condition that they seemed to have been born again.

Every physician encounters deplorable

cases of children three and four years old whose diet consists almost exclusively of meat, simply because their perverted appetites demand that article. In such extreme instances the most severe measures are justifiable in order to resume the natural and healthful method of feeding, to save the child's health if not its very life. We should permit it to become genuinely hungry by withholding all meat, or even all food, until it will consent to recommence taking milk. We may aid the child to overcome any temporary repugnance to milk by making it as palatable as possible. It may be aerated in a milk-shake, beaten in a cream-whipper, flavored by oyster juice and renamed "oyster soup," seasoned with any harmless essence, or made warmer or cooler, as the child may prefer. But milk it should have in some form, or be allowed no food at all.

Variety is desirable, and even necessary, in the diet of all children; but in seeking variety we should never lose sight of the main principle — that milk should be the chief and frequent article of diet, and meat, if not wholly excluded, admitted only as an occasional and non-essential part in the diet of any child under six years of age. Many children reach that age in superb health and with fine physical development without having known the taste of meat. The little one will naturally tire of milk if he is always given plain milk, milk, milk, without any change. But milk with oatmeal, milk with hominy, milk with cracked wheat, with cracked corn, with rice, with baked apples, seem in infantile judgment quite different dishes. There are also the various cream soups, made up without butter or seasoning beyond the natural pinch of salt. This also we may vary with a number of articles not taken with milk, but served in a different course, or offered as a separate meal.

In a rational and healthful dietary for a child may be included those fruits and berries that are not seedy or of too coarse a grain, such as peaches, sweet apples, strawberries, oranges, and dates. People frequently declare that their children, who are allowed to eat large quantities of the decidedly injurious meat, cannot take the really wholesome, desirable fruit. But we may be assured that the child to whom fruit is injurious has had his digestive organs brought into an unnatural condition by unwholesome diet. It rarely happens that a healthy child cannot be educated to enjoy and digest a large quantity of fruit. One especially vigorous baby three years old took regularly a large saucer of stewed prunes after breakfast, an orange after dinner, baked apples after supper, and a fruit luncheon of raw apples and dates half way between meals, not only without injury, but with positively demonstrated advantage.

As for the young child's dislikes in the matter of food, they are, of course, only the outgrowth of errors of judgment on the part of the parents. If a child has never tasted meat, or candy, or cake, or pie, he naturally can neither desire nor demand these undesirable articles. If he has always had milk as his chief diet, it will be impossible for him to dislike it or refuse to take it, unless the milk fed to him has been at some time tainted or acid, and so has given him a temporary repugnance to its use.

The milk that takes the place of meat must not, however, be thin in fats. It should show in the lactometer—a simply graded test-tube, easily procurable at any druggist's—at least twenty per cent. of cream. If milk of this quality is not obtainable, cream must be added to bring it up to this proportion. It is also of advantage to have the milk sterilized before it is fed to the child, since this process not only guards the child, as it does the infant, against the danger of contagious diseases, but averts the danger of irritation in the intestines, and prevents all the various forms of stomach and bowel trouble which so frequently result from the use of raw milk.

CHAPTER XIII

CONTAGIOUS DISEASES

THE average parent is too apt to consider the milder contagious diseases of childhood simply as inconveniences, of only temporary detriment to their victim. But we must recognize the scientific fact that no disease ever leaves the physical system absolutely unimpaired. To this we must add the fact that with healthy children growth is constant, and that the arresting of that growth by any disease really diminishes, to just such a degree as it extends, the ultimate size and vigor of the child who suffers from the illness. Contagious diseases, however harmless they may seem, should never be knowingly incurred; for even their least injurious results are unknown quantities militating against the development of the child, while there is always risk of more serious manifestations whose evil consequences may extend through the whole life of the child, and seriously impair both its usefulness and happiness. Therefore it is only our plain duty to guard against contagious disease as long and as far as may be. This is now possible to an extent never before conceived of. We at present understand, to a degree at least, the nature of contagious diseases, and out of this knowledge we gain power to avoid or to abort the disease.

After the determination of the germ origin of contagious diseases, special experiments were instituted among bacteriologists to isolate the germ of each disease known or suspected to be contagious; and as the microscope revealed the fact that more diseases than had previously been suspected were of germ origin, and were contagious, a reclassification of diseases was based upon this discovery. The most important item of the reclassification was the removal of tuberculosis, or consumption, from the non-contagious to the contagious list, since that disease now claims more victims and causes a larger number of deaths than any other known.

The most important result of all this recent bacteriological investigation has been the knowledge that contagious diseases are not incurred unless two conditions simultaneously favor their inception. We must come in contact with the specific seed of any particular disease, and we must furnish within the body, under conditions favoring its activity, the proper soil for the propagation of the germ, or no contagion ensues. That is, we must have direct exposure to the disease germ, coinciding with such a debilitated condition of the system as shall render it susceptible to infection. Exposure to disease will not result in contagion if the system be in prime condition; nor will impaired physical vitality lead to the contraction of contagious disease except with direct exposure. Therefore it follows that if with rational vigilance we guard against both these enemies, of which we are seldom forced to encounter more than one at a time, we shall safely resist any ordinary danger of contagious disease.

With children the usual method leaves not only a loop-hole, but actually a large breach for the entrance of any chance infection. While adults are almost without exception offered only food that has been cooked, in which the germ life is therefore destroyed, many babies are fed entirely upon food which is literally swarming with microbes, any meal of which may contain germs of contagious disease and certainly will contain bacteria that will at least produce such irritation in the digestive tract as to induce a decided lowering of the vitality, thus making the child's system peculiarly susceptible to any contagion. Every drop of milk or water fed to a baby, and all the food given to a child, should be boiled or sterilized; and each child should be maintained in the best possible general health, not only for his present comfort, but also for his future protection against infection.

To determine just when a child is or is not in average good health is beyond the power of unprofessional skill, but periodic physical examinations by a competent physician will detect even slight departures from normal condition. When it is comprehended that such systematic examinations are both less trouble and less expense than the diseases which are thus anticipated and prevented, more of our children will be allowed the benefit of such preventive treatment. We have our valuable cows and our imported sheep frequently inspected; even our trees and vines are yearly pruned and trained; but our children rarely receive any scientific attention at all until the active presence of acute disease has impaired their health and threatened even their lives.

The most conspicuous phenomenon of childhood is growth. The lives of all living creatures are rhythmic alternations of growth and decay. In childhood growth predominates; in old age decay is gaining the ascendency. Now the physical growth of the child is dependent upon the appropriation by its body of the proper material from outside itself, and is limited by the ability or fitness of the body properly to assimilate these materials. The elements absorbed by the system are taken as food into the digestive organs or as air into the lungs. Presupposing a supply of pure air and an adequate provision of nourishing food, such as has been suggested in another chapter, the body's ability to utilize these materials marks its degree of health and its capacity for growth.

To utilize to the best advantage these outside materials, food and air, two sets of functions, assimilation and elimination, must properly balance each other. Of all the matter taken into the body through the lungs and stomach, a large proportion of it is quickly eliminated; much of it is given off through the skin and lungs in the form of moisture and gas, and part passes through the excretory system. But after maturity the amount of matter daily eliminated is necessarily quite equal to that directly appropriated, though not of course the same material, in the same period of time, since the extra amount appropriated in childhood for purposes of growth is no longer needed by the adult. A certain proportion, which has previously been assimilated, and has become an essential part of the physical frame, is later thrown off as effete matter, to give place to newer, fresher atoms, slowly but constantly forming out of the extraneous material regularly consumed by the body. Eating, breathing, and sleeping help the body in its reparatory process. Exercise and bathing assist in the equally essential process of destruction and elimination.

A mother may, unaided, decide upon the quality of the material furnished to the body of the child. She may be entirely competent to judge of the relative values of the food and air which her children receive, but medical skill alone can determine if this material is properly appropriated by the various organs of the body. Malnutrition, which usually precedes and inevitably invites disease, is frequently unsuspected until actual acute illness demonstrates its existence. Diminished respiratory power likewise precedes, sometimes for years, any recognized manifestation of chronic lung-trouble. The degree to which the respiratory power of the individual child varies from the normal cannot be measured except by professional skill; but a physican can, in ninety-nine cases out of one hundred, select by physical examination, months or even years before the manifestation of any active lung-trouble, the people who are most likely to be attacked by pulmonary disease. With children to be forewarned is to be forearmed. With them any temporary trouble easily develops into chronic disease, while, on the other hand, they are happily more responsive to the preventive treatment that would follow any first unsatisfactory indications.

CHAPTER XIV

VARIATION OF RULES

During the second year of life a child is usually disposed to shorten its day naps. Two naps gradually give way to one, and even that one is, with increasing years, increasingly difficult to maintain. The careful tactics that will induce sleep in that comfortable animal, a well-nourished baby, no longer suffice with an older child, who often persistently resists the inclination to sleep. It is the children of nervous temperament, who really need the greatest amount of rest, with whom it is most difficult to continue the daily naps after the second and third years. The incessant physical and mental activity of the growing child puts great strain on all its powers—such strain as even the most energetic adult would be incapable of enduring. If this activity is continued unbroken for the twelve or fourteen hours of a child's day, it cannot but become a terrible drain upon the constitution of even the most vigorous child. The maintenance of regularity in day naps with all children under six years of age is, therefore, important enough to merit especial effort.

The best rule at which to aim is of course to follow the practice of the first year, and at the regular time for sleep, either by day or night, to place the child awake in its bed, make it thoroughly comfortable, darken the room, and leave it to fall asleep by itself. But as the rule is less important than the object for which it was made, if we cannot by strictly adhering to it accomplish our purpose, we must then adopt such modifications as appear necessary to induce the child to sleep. An active child cannot be snatched from the floor and, after the brief interval required for the process of undressing, summarily deposited in its bed with any certainty that it will not continue its play from that point of vantage with as great hilarity as ever. Left to itself, the child becomes every minute more wide awake, more nervously active, and less in condition for a restful night's sleep; whereas if it is rubbed or bathed, and taken into the arms for a quiet story or soothing lullaby, it may afterwards be deposited in bed, if not asleep, at least so quieted and drowsy that sleep inevitably results within a few minutes. If a child who requires an hour to fall asleep by itself will drop off in five minutes if rocked or sung to, it is manifestly better that the mother should lose her five minutes of time and the child gain its extra hour of sleep. It should always, however, be remembered that such aids to sleep are exceptions to or modifications of the ideal system, and are made to meet the personal idiosyncrasies of the individual child. In their adoption we must not lose sight of the general rule that it is far better for the average child to fall asleep by itself, in the quiet darkness of its own room.

A continuation of the regular habit of perfect rest once during the day, even, where sleep cannot be induced, a half-hour of absolute relaxation of the muscles and rest for the eye, ear, and tongue, is of the greatest advantage. To this habit many older people undoubtedly owe the blessing of vigorous health coexisting with the power of continued and exhausting brain-work.

CHAPTER XV

THE NURSERY

In the development of a child, vigorous exercise must rank in importance with nourishing food, pure air, and sufficient sleep. The last generation of children was not so fortunate as is the present in the matter either of good food or pure air, since the scientific importance of these materials for the building up of the body was not then so carefully regarded. But in the matter of exercise the children of to-day are increasingly unfortunate. The tendency of population has for many years been most decidedly towards condensation; and the limitation of space in our large cities, with the consequent overcrowding of nearly all our city houses, tends to confine the children to one room.

This room, often, and indeed generally, overcrowded with furniture, permits the children no liberty to indulge in the most natural form of exercise—running.

What is needed is a change in our estimate of relative values. We must come to realize the truth that the first few years of a child's life are the most important in their bearing upon his ultimate physical condition and mental attainment. To the children should therefore be devoted not the worst but the best room in the house; best not in elaborateness of furniture nor complications of decoration, but best in size and position; best in receiving the greatest quantity of sunlight and fresh air. What furniture is absolutely necessary—and nothing should be permitted that can be omitted — should be placed around the room against the wall. No furniture, unless it be an unobtrusive chair, should obstruct the centre of the room. All necessary seats should be in the shape of low couches or divans well cushioned. These possess many advantages over any form of chair. While they furnish room for the nurse or mother to sit, they make a capital place where the children may play, from which or upon which the tumbles are never serious, and on which the child may lounge or play when it could not be induced to sit in a chair.

The floor should be always bare, of either painted or hard wood, and covered in the centre by a thick warm rug. It is often urged that bare floors, although much to be preferred in point of cleanliness, are undesirable for a child who creeps or plays most of the time upon the floor. If a baby is learning to creep in cold weather, it is not, however, necessary, and certainly not desirable, that it should be allowed to creep upon the floor at all. The value of creeping bears no relation to the distance through which the child propels itself. Creeping is simply the preliminary exercise by which a child strengthens its limbs for the initial effort to walk. It gets just as much and as valuable exercise by crawling back and forth over a properly protected surface three feet by five as it can by sweeping a floor fifteen feet by twenty. It saves infinite trouble with a creeping child, and protects it against many colds and much dirt, if it is confined in a pen placed in one corner of the room; or, better still, the child may be raised from the floor by placing him on some low couch surrounded with a railing. Such a pen, while it may be contrived easily and without much expense, may also be designed so elegantly as to be really an ornament to any room in the house. In this enclosure a baby may be placed during the months from the period when he begins to creep until such time as he has learned to walk with certainty and vigor. By means of the sides of the pen he is soon able to raise himself to his feet, and by clutching its firm rail he easily learns to walk round its circumference, which to him seems endless. With a few simple playthings for company inside the rail, and with a friendly face and voice outside but within sight and hearing, the child, during this usually most troublesome period of its young life, becomes simply no trouble at all, but grows and thrives to the extent of its power, and demonstrates conclusively that it is absolutely unnecessary for a creeping baby to undertake the dangerous navigation of the nursery floor.

Older children can be taught to choose, when playing upon the floor, the part that is protected by the rug. But the average child sits on the floor by far too great a portion of the time. It is very easy, by a little forethought, to counteract this tendency by providing a table, such as is used in the kindergarten. Even a plain cutting-table will serve the purpose. Sitting or standing beside this, the child will find upon its limited surface sufficient room to create a world of interest. By the force of his vivid imagination it becomes successively a complication of railroad tracks, a field of exciting battles, a barn-yard, or Mount Ararat disgorging the inhabitants of the Ark. By this provision of a table or tables there is less conflict and misunderstanding, even where several children are engaged in play, than is possible by the indiscriminate use of the floor surface, since each child may enjoy exclusive right to his own little table or definite portion of table, and within its limited space rule as undisputed monarch.

If in addition to the nursery the older children can, during portions of the day at least, have the use of some adjoining room or hall which, with the nursery, furnishes space enough for a good run, it is of great advantage, especially when inclement weather or temporary indisposition prevents them from enjoying the usual out-of-door exercise.

Throughout the children's quarters the furnishing should be extremely simple. First, in order that it may occupy as little room as possible, and, next, that it need not be too valuable to endure the rough usuage it is certain to encounter at the hands of the nursery vandals. Children should not be encouraged or even permitted to indulge unduly their natural instinct for destruction; they must not, on the other hand, be continually worried by warnings not to touch this, or injure that, or break the other. Every article in the child's room should be there for his particular convenience and enjoyment, and he should be allowed its full, free use, being taught, meanwhile, the difference between the use and the abuse of his own property. Neither should he be reproved or punished for any accidental or occasional injury to the articles he handles. The muscles of the little fingers are not yet firm; cerebral development is not yet sufficiently co-ordinated to control their action. And therefore, while it is proper to express sorrow or regret at any accidental destruction, the child should not be alarmed or punished for an occurrence for which he was in nowise responsible.

CHAPTER XVI

TO AVOID SELF-CONSCIOUSNESS

It is a vital error to magnify in the mind of a child the importance of things; neither clothes, nor furniture, nor ornaments, nor playthings should be magnified to the detriment of the child's happiness or health. We of the adult world, who are in a sense slaves to the things we think we own, should at least save the children from a too early and oppressive sense of subordination to inanimate objects. The nursery may be artistic and comfortable, and still contain nothing that fills or litters it, and nothing that is easily injured or destroyed. A child's playthings may be sufficiently bountiful to satisfy all his natural demands for amusement without including anything that will easily tear or break, or in whose use he must constantly be warned to be careful. We ourselves may reduce the occasions of reprimand and warning to such an extent that we shall neither weary nor tire him. Windows should be guarded by strong bars, stairways protected by swinging gates, the open fire shielded by a screen, rocking-chairs and light furniture with sharp corners banished from the nursery; and then the little one may safely be allowed to seek his own amusement and make his own investigation throughout the whole limit of his domain, free from the constant repetition of "Don't" and "You must not," which is very wearing to his temper, and totally incompatible with his best mental and physical development.

While the child should not be interrupted nor hampered in his childish occupations, neither should he, on the other hand, be spoiled by too much entertaining or assistance from older people. His own methods of investigation and his natural instincts in seeking certain kinds of amusement are the least exciting, and at the same time the most instructive, because they are the most natural. Children who are constantly nagged at, for that very reason require constant amusement, while children who are rationally and judiciously neglected will soon learn to entertain themselves.

In the matter of dress, the ideal condition with the child, as indeed it should be with the adult, is to devise such a costume as shall, first of all, not make the wearer self-conscious. He should be as unconscious of his clothes as an animal is of its fur, or a bird of its feathers. With little children the only element necessary to contribute to this result is to provide an abundance of dresses of uniform quality. The mind of a young child is incapable of remembering. We give ourselves and it much trouble from failure to appreciate this fact. The young mind is intensely receptive, and absorbs an infinite number of vivid and varying impressions;

but in its undeveloped condition it is incapable of consecutive or prolonged effort. The very structure of its brain precludes the possibility of its keeping in mind our warnings and injunctions. No child can remember to keep clean. If we enjoin upon it a hundred times a day not to soil its dress, we nag just so many times and produce no permanent effect whatever. It should therefore be provided with such an abundance of dresses or aprons that we may remove entirely from its overburdened mind any responsibility for keeping clean, and transfer the duty to the nurse, who should discharge it by frequent renewing of the simple outer garment exposed to the dirt.

With an older child, in whom the power of comparison is developed, we need to take other precautions to prevent self-consciousness, which so frequently grows out of parental error in juvenile costume. All our preconceived notions as to artistic regulations or the requirements of good sense should be waived in favor of general usage, at least for the outer garment, which is most in evidence, so that the child may not by any comment or criticism on the part of his companions be made conscious of singularity in the cut or quality of his clothing. His garments should not be elegant, or shabby, or eccentric, as judged by the average standard of his playfellows. No superabundance of riches should tempt a mother to dress her child conspicuously better than its companions. If the stress of poverty renders it absolutely necessary that a child should be dressed less well than his fellows, the reason and the necessity for such digression should be clearly and simply explained as soon as his comprehension is adequate to such reasoning. The circumstances should be presented to him as making the highest claim for self-sacrifice on the ground of the good of his family. Then his shabby dress will help him towards self-discipline rather than self-concentration.

Many a child artistically and sensibly dressed yet suffers tortures of which the adult mind can hardly conceive, simply because his costume differs from that of his comrades. Causes seemingly unimportant can yet cruelly wound a nature over-sensitive to ridicule. To yield to a child in a matter of which he cannot be the best judge seems to many parents folly, whose outcome is to pamper and spoil him. But, on the other hand, we must consider that the cut and color of a child's garment are far more important to him than they can be to us, and that as long as the under-garments conform to hygienic rules, the outside garments can usually be modified without any sacrifice of health or discipline—certainly without any detriment to the health of the child or the dignity of the parent. It is not merely a question of the child's will against our will. The principle involved goes deeper, touching the dangerous undercurrents of a child's character. For intense self-consciousness — easily incurred in childhood, almost ineradicable in manhood—is always intensely obnoxious, and cannot but be fatal to the best development of its unhappy victim.

CHAPTER XVII

THE NURSERY-MAID

WITHIN the past ten years a great interest has been aroused in the scientific training and development of young children. This is partly due to the general spread of intelligence concerning the importance of the early years of life, partly to intelligent study of child-life by the kindergarten enthusiasts, and somewhat also to the great increase of college-bred women in the ranks of mothers, physicians, and teachers. But this healthful activity has not yet had time to crystallize to any great extent into practical rules, nor has it, so far, altered the fact that American children of the well-to-do classes are less well cared for than in England. They do not enjoy such average good health, and their manners are proverbially as bad as it is possible for them to be.

Though the statement may at first seem unfounded, some little investigation will yet prove that this condition arises from the fact that American children receive too much attention and too little care. They also enter at too young an age into the stimulating, unwholesome excitement of the life of the adult portion of the family. In England the children, not only of the upper classes, but even of the middle classes in moderate circumstances, are provided with their own apartments and watched by a special care-taker; and the routine of their daily life is suited to their immature years, as favorable results demonstrate. The children of an English family are by no means neglected by the mother. She personally formulates all the rules for their daily life, is a frequent visitor to the nursery, presides at many of their meals, receives from the nurse daily reports of their variation in health or departure from good conduct. She has a competent, well-trained nurse, sometimes two or three, to execute in detail her requirements. In America, in a family of equal social and financial standing, the mother either has no regular nurse at all, or, for economy's sake, puts up with an incompetent one; while among the wealthier classes, where one or more expensive nurses are secured, the mother almost universally abandons the supervision of her children, and practically leaves them to the less intelligent care of these women, whose personal peculiarities and habits are and must remain an unknown quantity to her, while she makes only a brief daily visit to the children's quarters.

It sometimes happens that a young mother is conscious, in the care of her children, of real inferiority to her nurse. But it is possible, and ought to be considered important, for every mother to learn at least as much as an ordinary uneducated nursery-maid can know. If she cares for her child herself during the first six months of its life, she will already start the superior of the nurse in that she has attained some comprehension of the child's peculiarities and individual tendencies. During the period of seclusion preceding the birth of the child she will find preliminary reading an opportune and enjoyable occupation. Books and magazines on the care of children are multiplying in the land, and no expectant mother is now without means of obtaining the necessary information.

In securing a nursery-maid it is usually better for a mother to engage a young girl of superior intelligence who is willing to learn, rather than to attempt to obtain an experienced nurse who already fancies she knows everything, and consequently ignores any instruction not consistent with her ancient code of nursery routine. Moreover, "experience" too often means a knowledge of how to keep a child quiet with drugs, to frighten him into submission with fearful stories or dire threats, to spoil his temper with needless opposition, or to correct his faults by punishment. The new and infinitely valuable scientific methods for the care of children are absolutely incomprehensible to the old and "experienced" nurse. Incomprehensible they may also be to the young and inexperienced nursery-maid; but her very ignorance makes her willing to carry out even that which she does not and cannot thoroughly comprehend, and which it is not important that she should understand from its scientific side, if the mother will direct and supervise the daily routine until it has become force of habit with both nurse and child. There are absolutely no experienced nurses, measured by the new methods of caring for children. Acknowledged inexperience is therefore our safest refuge.

The most important qualifications in choosing an applicant for this systematic training as a nurse are good health, personal cleanliness, fondness for children, good temper, a cheerful disposition, a desire to follow instructions, and absolute truthfulness. Some of these qualities can be ascertained only upon trial; but as the combination of all the qualities necessary is not unreasonable or unusual, we may at least hope to find them more often than we have been able to secure the more exceptional and really less desirable quality—long experience.

Such a routine of daily life as that which these articles advise, which has been tried in hundreds of cases and proved to produce most desirable results, may certainly be easily understood by the average mother, and by her put into practice with moderate exercise of personal care. It may be taught to any intelligent nursery-maid, and after a few months' practice carried out by her in detail, with a careful daily report of its results to the mother, who shall give only a general supervision.

We believe that the children who are totally neglected by the mother, and those who are cared for solely by her, are both, on an average, unfortunate. No child can attain his maximum physical vigor or best mental development without the supervision which maternal affection and superior intelligence can give. On the other hand, the mother is necessarily absorbed in many interests; her husband, her house, her friends, her church, her relatives, occupy much of her time and attention, so that in her effort to do too much for her children she really does by far too little. Some of the infinite details necessary to the health and comfort of her children are inevitably neglected. Buttons can be sewed on, dresses let down, faces washed, clothes kept in good condition, meals properly served; dressing, undressing, naps, and exercise can be as well and better attended to by the nurse than by the mother. The definite amount of strength and the limited number of hours which a mother has to give to her children each day may better be reserved for more important items than buttons and dirty hands. These items are not the most important; but in the pressure of every-day life the material details in the care of children are often so absorbing that they consume literally all the mother's time and thought, so that she has neither energy nor opportunity for the higher duties which touch on mental and moral development. She has no time for her own personal development, for companionship with her husband, for the recreation of social life, nor indeed for anything but buttons and dirty faces.

Even where the absence of a nursery-maid has, for reasons of economy, seemed necessary, her employment would become possible if its true value as bearing upon the ultimate health and happiness of the family were properly appreciated. It is simply a question of the relative importance of the various expenses. In America many a woman thinks she must have diamonds and costly costumes, even though she is forced, in consequence, to manage without a nursery-maid. In England a mother in moderate circumstances thinks she must have a nursery-maid, even though she never owns a diamond nor wears anything better than a home-made stuff dress.

Many American mothers cherish the misguided sentimental impression that no one can give the children such good care as themselves. But no mother can give her children really good care and do anything else, without absolute demoralization of her own health and complete arresting of all personal mental growth.

THE END

www.ingramcontent.com/pod-product-compliance
Lightning Source LLC
Chambersburg PA
CBHW030303170426
43202CB00009B/851